Busy Bugs

BUTTERFLIES

By Bray Jacobson

Please visit our website, www.garethstevens.com. For a free color catalog of all our high-quality books, call toll free 1-800-542-2595 or fax 1-877-542-2596.

Library of Congress Cataloging-in-Publication Data
Names: Jacobson, Bray, author.
Title: Butterflies / Bray Jacobson.
Description: New York : Gareth Stevens Publishing, [2022] | Series: Busy bugs | Includes index.
Identifiers: LCCN 2020006215 | ISBN 9781538263297 (paperback) | ISBN 9781538263303 (6 Pack)| ISBN 9781538263310 (library binding) | ISBN 9781538263327 (ebook)
Subjects: LCSH: Butterflies–Juvenile literature.
Classification: LCC QL544.2 .J34 2022 | DDC 595.78/9–dc23
LC record available at https://lccn.loc.gov/2020006215

First Edition

Published in 2022 by
Gareth Stevens Publishing
111 East 14th Street, Suite 349
New York, NY 10003

Editor: Kristen Nelson
Designer: Katelyn E. Reynolds

Photo credits: Cover, p. 1 YURI CORTEZ/AFP via Getty Images; pp. 5, 23 borchee/E+/Getty Images; pp. 7, 24 (wings) Garry Gay/ Photographer's Choice / Getty Images Plus; p. 9 Kelly Kalhoefer/The Image Bank / Getty Images Plus; p. 11 Louise Docker Sydney Australia/Moment/Getty Images; p. 13 Robert Landau/ Corbis Documentary/Getty Images; pp. 15, 24 (eggs) Wild Horizon/Universal Images Group via Getty Images; p. 17 teptong/ iStock / Getty Images Plus; pp. 19, 24 (pupa) stanley45/E+/Getty Images; p. 21 Willowpix/E+/ Getty Images.

Printed in the United States of America

CPSIA compliance information: Batch #CSGS22: For further information contact Gareth Stevens, New York, New York at 1-800-542-2595.

Contents

Butterflies are busy!
They come out in daytime.

Most have two pairs of wings.

They have six legs.

They eat plants.
They drink from flowers.

They may fly a long way.

They lay eggs.

Babies look like worms.

They rest as a pupa.
Their bodies change.

They become butterflies!

They are bright colors.

Words to Know

eggs

pupa

wings

Index